JAPANESE FOLKHOUSES
NORMAN F CARVER JR

Books by Norman F. Carver, Jr.

FORM AND SPACE OF JAPANESE ARCHITECTURE
SILENT CITIES, MEXICO AND THE MAYA (1965, rev. edition 1986)
ITALIAN HILLTOWNS
IBERIAN VILLAGES

NORMAN F CARVER JR

JAPANESE
FOLKHOUSES

DOCUMAN PRESS LTD

ACKNOWLEDGMENTS

These photographs of folkhouses and villages were made during Fulbright grants to Japan more than 20 and 30 years ago. I am delighted to have them finally displayed especially since, with the extraordinary changes in Japan, many of these photographs could no longer be made.

Those were wonderful times in Japan—each day filled with new and exciting discoveries. They would not have been possible without the friendly cooperation of the Japanese people, who graciously opened their homes to us. I must also thank the Fulbright staff from those early days who helped make our stay not only pleasant but productive; our fellow grantees for their support and companionship; Dr. K. Horiuchi in Kyoto for his introductions into the folk movement; Dr. Yas Kuno and his family for their friendship and for sharing their house in Kyoto; and Senator William Fulbright for his foresight in establishing this exchange program which has significantly enhanced thousands of lives around the world.

My wife Joan was invaluable in the final selection of photographs and editing the text. Without her help the book would never have been finished.

©1984, 1987 Norman F. Carver, Jr.
First Printing 1984
Second Printing 1987

Documan Press, Ltd.
Post Office Box 387
Kalamazoo, Michigan 49005

ISBN: 0-932076-04-1 Cloth
ISBN: 0-932076-05-X Paper

Designed by Norman F. Carver, Jr.
Printed by EPI, Inc.
Battle Creek, Michigan
Printed in the United States

To my family—my wife Joan and my children, Norman III and Cristina, with whom I shared many of the adventures and delights in Japan.

CONTENTS

PREFACE	6
COMMENTARY:	
JAPAN AND THE JAPANESE	7
ANCIENT ARCHETYPES	10
FOLKHOUSES—MINKA	42
PHOTOGRAPHS:	
ANCIENT ARCHETYPES	16
KYOTO AND THE WEST	49
THE FAR COUNTRY	145
BIBLIOGRAPHY	196
INDEX	197
MAP	199
PHOTOGRAPHIC NOTES	200

PREFACE

This book really began more than thirty five years ago, when I was shipped off to Japan by the United States Army. Fortunately stationed in Kyoto, and fascinated by the glories of that ancient capital, I returned five years later to study and photograph Japanese architecture on a Fulbright grant.

My wife and I had just finished our architectural studies so the Japanese experience, at a crucial time in our careers, made an indelible impression upon us. We found an architecture clean-lined and ordered, yet expressive and spatially exhilarating. After a thorough dose of Miesian asceticism at Yale, it was like a breath of fresh air. Japanese architecture, of all the world architectures I have explored, remains the most potent influence in my own designs.

Those early photographs, initially made without any intention of publication, allowed me to discover for myself what was so deeply affecting about Japanese forms, and to express that feeling through very personal photographs. Eventually, out of my second stay in Japan came a book, FORM AND SPACE OF JAPANESE ARCHITECTURE, which, much to my surprise, remained in print for nearly 25 years. But it only scratched the surface and in the 1960's I returned to concentrate on photographing the folkhouses and villages. Completion of this book was delayed, however, when a trip to the Mediterranean compelled me to begin documenting that rapidly disappearing indigenous world.

At a time when architecture appears lost in the self-indulgent and irrelevant aesthetic play of 'post-modern', my studies have convinced me that vernacular architecture, and especially the Japanese folkhouse with its striking affinity for basic tenets of modern architecture, provides desperately needed insights into the fundamental connections between man, society, nature, and architecture.

JAPAN AND THE JAPANESE

Ever since its revelation to the world in the late nineteenth century Japanese architecture has been a powerful influence on Western design. The main influence on domestic architecture stemmed from the sophisticated houses and tea houses of the upper-class. Traditional folkhouses, except in works like E. S. Morse's book[1] in the 1880's, were largely neglected by both Japanese and foreign scholars. During the last few decades, however, just as they began rapidly disappearing, folkhouses suddenly have been hailed as the embodiment of Japanese architecture's most fundamental values, if not its highest aspirations.

Folkhouses

Most histories of architecture have ignored the traditional common house; yet it is among man's most complex and ubiquitous creations —a product of a physical and emotional relationship with human existence that has been constant, intimate, and profound.

As shelter, folkhouses were essential to survival by moderating the extremes of climate, by keeping the terrors of the outside world at bay, and by providing the spaces that made life and work possible in an uncertain world.

But the folkhouse provided more than shelter; it was also a powerful, emotional symbol—a symbol, for example, of the inhabitants' wealth and status in the community (think how often today we judge someone by a view of his house). Privately the house symbolized family life—and, no matter how humble or plain, would remain forever 'home' to those who grew-up there. As part of this symbolic role folkhouses were the principal outlet for aesthetic expression, the focus of most of man's, and especially woman's, creative efforts.

Many languages reflect the intimate relationship between man and his house. In Japanese, for example, 'kanai' means wife as well as inner room, and a word for house, 'uchi', can also refer to oneself.

The combined functional and symbolic roles make the folkhouse, more than any other man-made object, a true reflection of its physical and cultural context.

Nowhere is this truer than in Japan. To appreciate the fit of the Japanese folkhouse to Japanese life some understanding of the physical, cultural, and historical background is necessary. In brief, Japan combined a physical environment of limited space, mild climate, plentiful rainfall and abundant forests with a rigid social structure under strong central authority which grew out of an early history of fierce clan rivalries. A native Shinto reverence for nature overlaid by Buddhist asceticism led to cultural ideals of frugality, and physical self-denial. Also significant was Japan's cultural and geographic isolation from the rest of the world for most of its existence.

Architecturally these circumstances resulted in an open and flexible wood architecture which was both surprisingly uniform throughout Japan and richly varied region to region; a folk architecture which is unique and yet vaguely reminiscent of folk forms in other, distant cultures with similar environments (the Alps, for example). In other words, Japanese folk architecture exemplifies the idea of folk architecture as the pragmatic resolution of both cultural and environmental forces in which these forces modify each other over time to produce a high degree of integration between house style and life style.

The Land and Climate

The four large Japanese islands, formed by a series of volcanic peaks rising from the ocean floor, stretch for some 1200 miles along the Asian mainland. The climate ranges from semi-tropical in the south to semi-Siberian in the north—similar to the American east coast or the European Continent. Typical of a monsoon country, heavy rains fall daily for several weeks during late spring. Plentiful rainfall and the temperate climate of most of the islands combined with the limited growing space of the mountainous land created an ideal setting for the rice-based economy. The islands' other characteristics are less sanguine. Destructive typhoons and earthquakes occur frequently—terrifying forces in the back of every Japanese mind.

The most influential climate architecturally, is that of the southwest part of the main island, Honshu, along the Inland Sea. Here in prehistory the essential character of Japanese culture and architecture originated. Winters in this region are mild and damp and summers, ushered in by the monsoons of June, are oppressively hot and humid. Early houses emphasized summer comfort over winter warmth not only because of their origins in this predominately warm climate, but also, as some evidence suggests, from prehistoric connections with tropical southeast Asia. This original emphasis had a profound and lasting effect on house form as well as social mores. It meant light-weight, movable interior and exterior walls for maximum ventilation causing a lack of personal privacy. It meant large sheltering roofs to keep out wind-driven rain and sun which became a dominant design element and symbol. It meant a strong post and beam structural system to resist typhoons and earthquakes, to accommodate the movable walls, and to support the massive roofs. And it meant little effort to heat interiors in winter forcing a somewhat stoic lifestyle at least among the lower classes.

Surprisingly, the original semi-tropical style spread easily to less hospitable climates in the islands with only minor variations in form. This stylistic migration ran counter to a basic characteristic of folk architecture in general—precise adaptation to local climate. Methods for increased winter comfort were certainly available (the ancient Korean floor heating system, for example), but the Japanese failed to incorporate them into their building tradition. Apparently the extreme uniformity of Japanese culture combined with the idealization of the simple frugal life and the tendency to stoicism and self-denial in the Japanese character overrode more practical considerations.

Whether or not such stoicism was innate or bred in generations of drafty bed rooms, it parallels a relationship between house form and culture also imposed by the openness of the Japanese house—the elaborate social conventions that dealt with a lack of personal privacy.

The Japanese

In observing other cultures, we must, of course, always be outsiders, although this may allow us to perceive qualities long since obscure within the culture itself. The Japanese regard themselves and their society as especially inscrutable to foreigners—believing that no outsider can ever really understand them. Many Japanese have made this observation —often the same person who returned from a month's visit to New York and wrote a book explaining all about America.

Centuries of isolation gave the Japanese both a sense of superiority and insecurity when dealing with alien cultures. The feeling is not recent. Very early in their history they discovered a Chinese civilization more sophisticated than their own from which they proceeded to borrow administrative techniques, religion, art and architecture—importing ideas and advisors wholesale in a largely successful effort to 'catch up'. Again in the mid-nineteenth century after a long period of self-imposed isolation, their renewed contacts with the West shocked them into a new wave of imported technology and ideas. And the post-war transformation of Japan is a phenomenon felt world-wide.

The Japanese opinion of themselves and their culture as exceptional stems largely from their island existence. Unconquered by foreign armies, they were remote enough from the rest of the world for a singular culture to develop, yet close enough for occasional stimulation from more advanced nations on the Asian mainland.

Origins

The precise origins of the Japanese culture and the Japanese people are unknown. Most evidence suggests pre-historic migrations from Korea, Mongolia, and Southeast Asia—disparate groups that somehow coalesced into the distinctive Japanese race.

For most of their early history the Japanese lived as isolated, feuding tribes in the southwestern portion of the islands, gradually displacing the Ainu aborigines. In about the first century the legendary Emperor Jimmu forcibly unified the diverse clans to become the founder of the Japanese nation and the Imperial Family. Even in those distant times Imperial power was impressive as evidenced by the huge mound tombs they left behind—some of the largest such structures in the world. Much of Japan's later history alternated between periods of peace and clan wars swirling about the ever-changing fortunes of this same Imperial Family.

The rivalry affected Japan's architectural development by causing regional isolation for long periods that led to development of local variations, although frequent attempts at unification and authoritarian control fostered homogeneity. But the compactness of the islands, their isolation and freedom from foreign conquest gave the Japanese an overriding sense of unity and the luxury to selectively pursue outside ideas and, even more important, the time to transform them into something distinctly Japanese.

An architectural example of this transformation is seen by comparing the Buddhist temples of China, which were so influential in Japan after the sixth century, with Japanese temples of the seventh century, such as Horyuji (which still stands). The richly decorated buildings and exotically curved roofs of the Chinese were simplified to fit Japanese sensibilities—sensibilities most clearly seen in the straight lines and unadorned surfaces of Ise (ee-say) Shrine and still eloquently embodied in the Japanese folkhouse.

ANCIENT ARCHETYPES

If, as seems likely, some of the Japanese originated in Southeast Asia or the Pacific Islands, they undoubtedly brought with them a house type still in use today in that region—a raised platform on poles covered by a large thatched roof. Although for much of Japan such elevated houses would be impractical in the winter, similar storehouses were widely used in ancient times to protect food from the damp and vermin.

Another primitive house type, whether indigenous or not is unknown, consisted of a circular or squared off pit a few feet deep with a pounded earth floor covered with a thatched roof. The roof poles were dug into the ground around the pit and lashed together at the peak. Large houses of this type had supporting posts in the center. Despite its primitive simplicity, this form established many of the basic elements and construction methods of the Japanese folkhouse. For example, the crossed ridge poles were the origin of the projecting rafters in the ancient shrines of Ise (17ff) and of similar projections on roofs near Kyoto (93) and still used for rice drying racks (167).

The materials readily available determine the fundamental character of folk architecture—in the sense that they either encourage or discourage certain forms and techniques. The abundance of wood profoundly influenced the course of all Japanese architecture for it was the primary material for every building style (except, of course, castles). As a consequence Japan is one of the few places where wood has been the dominant building material throughout history. Other cultures often evolved from

Reconstruction of prehistoric pit dwelling and raised storehouse-granary at Toro, near Shizuoka.

wood to masonry buildings or combined the two materials. In Japan, not only were the earliest primitive huts built of wood, but also the most sophisticated palaces and temples of recent history—resulting in an architecture of unparalleled unity over type and time. So ingrained is this love of wood, even today it is the material of choice when conditions allow.

Of early houses on raised poles there are no remains, only a few crude lines on bronze mirrors showing the major elements—posts set in the ground, a ladder leading to the raised floor, and separate posts at the ends supporting the center ridge—the same elements still seen at the Ise Shrines. Although it may seem unusual in a book on folkhouses to include religious monuments, the Shinto shrines of Ise, Izumo, Shinmei-gu, and the Shosoin storehouse provide the unusual opportunity to study existing examples of ancient building forms and techniques—to examine the very archetypes of the Japanese folkhouse.

Ise Shrine

In the year 680 AD a decision was made that profoundly influenced the future course of Japanese architecture. The Emperor Sujin directed his daughter to establish somewhere outside the capital, a central Shinto shrine dedicated to the goddess Amaterasu, mythical founder of the Japanese nation. Following a long search, the princess selected a forest in the remote province of Ise. Repository for Japan's holiest treasure, the sacred mirror, the shrine was built appropriately enough in the form of an ancient storehouse, similar to those elevated houses cast on ancient bronzes. It is a building type still found in Indonesia, Sumatra, and the Philippines, and on a few small Japanese islands such as Oshima. But, as my earlier book IBERIAN VILLAGES shows, this form is not unique to Asia—it is a common solution to the problem of protecting food and family treasures in Europe and elsewhere. Therefore similarity alone cannot be conclusive evidence of Ise's southern Asian origins.

Because of Ise's sacred role and its inherent relationship to the Imperial family, because of its remoteness from the changing fashions of the capital, and especially by its periodic renewal, Ise perpetuated the elemental purity of ancient building forms and became the eternal model against which all subsequent Japanese architecture was consciously or unconsciously measured.

The selection of this modest, expressive structure of plain undecorated surfaces as a focus of national reverence did much to determine the course of Japanese architecture for the next fourteen hundred years. It is impossible to explore the nature of Japanese architecture, especially Japanese folkhouses, without some understanding of these ancient forms.

In evaluating the influence of Ise it is important to remember that the shrines are the repository of the most ancient relics of Japanese history, a place revered and visited by every Japanese at least once in his lifetime. Ise represents his roots and the soul of his nation; what he experiences there becomes a part of him, carried with him the rest of his life.

This sense of sacred history is on the mind of each Japanese when he visits Ise. In the

Overall view of the Naiku Shrine complex at Ise from an 18th century travel guide. The slightly different architectural features from today's shrine are probably artistic license, but the woodcut gives a good impression of the forest setting and the throngs of pilgrims.

prewar years when the cult of the Emperor brought Ise new prominence, the Japanese made their pilgrimages with a mixture of awe and dread. Today, even for a Westerner, crossing the bridge over the Isuzu River into the ancient forest surrounding Ise causes a mysterious feeling of anticipation. Few places can evoke this reaction. Considering its modest scale and an overwhelming sense of nature, Ise may be unique. More often we are awed by sheer size or opulence as at the pyramids of Egypt and the New World, the temples of Angkor, or the Acropolis of Athens.

Approaching by a simple white gravel path through huge cryptomeria trees, one walks quietly on the gravel and speaks, if at all, in a hushed voice. A feeling of some primeval sanctuary far from the outside world pervades the forest. A glimpse of the shrine's bold roof forms accentuated by the crossed rafters signals arrival. There is something strangely moving about these soaring forms—their primitive simplicity evokes some ancient memory in man, marking this place as special and mystical (19).

Fences enclose the white gravel compound, clean and level, set in the midst of the forest. Arranged on north-south axes the main building or shoden stands between two smaller 'treasuries' built without the surrounding veranda or grand stair of the main building (22). Today the visitor to Ise is permitted only through the gate in the first fence, but can glimpse the shoden over the fences or through the inner gates.

Ise has two major shrine complexes and several smaller subsidiary shrines. The two major shrines are the Naiku or Inner Shrine, the more important original shrine at Ise—and the Geku or Outer Shrine at another, less dramatic site on the opposite side of the river.

The walls of the shrines, now of smooth planks set into grooves in the columns, were originally log-cabin style similar to the grand Shosoin in Nara (36-39). One small treasury at the Geku still preserves this method of interlocking boards at the corners as do some of the outer shrines.

One amazing aspect of Ise has been its regular rebuilding every 19 or 20 years since the year 690 (except for a few intervals, the longest, 123 years from 1462 to 1585, during great national strife). The current buildings, completed in 1973, are the sixtieth generation.

Except for minor variations and inevitable refinements, the shrines faithfully preserve the essence of the pre-historic forms—in part because of their sacredness; in part because for centuries rebuilding was entrusted to one clan of carpenters who passed their skills and knowledge from generation to generation; and in part because construction on alternating sites provided a model to discourage tampering with tradition.

But rebuilding also allowed the gradual introduction of refinements or enhancements to the grandeur of the shrine. Over the years some of the original rugged simplicity has disappeared, replaced by meticulous detailing, extreme precision, and highly finished surfaces (24, 25). Other additions include metal decoration on the ends of beams and on the balustrade. Most disturbing architecturally is the separate, temporary-looking roof over the front stairway which, though sheltering the steps and the priests during ceremonies, diminishes the impact of the shoden and its huge pillars thrusting from the ground. The intended effect, though much reduced in scale, can be seen in the adjacent treasury buildings in the main compound and in minor shrines at the site (22).

Ancient houses also used posts buried directly in the ground. Although this greatly strengthened the structural frame, the problem of rotting timbers caused abandonment of the practice. At Ise, where the buildings are periodically renewed, the huge pillars are still set deep in the earth—enhancing the sense of strength and connectedness to nature inherent in Shinto religion.

Built alternately on adjacent sites, after dedication of the new shrine, the old is dismantled. On the cleared site the only location marked is that of the sacred central pillar—reminiscent of the honor still paid to the large central post in many folkhouses. In the 1950's I visited Ise during the transition year when both sites were occupied. One compound of new tawny wood, rather stark and somehow less impressive in its newness, contrasted with the old compound of weathered grey-green wood and moss-covered thatch which, though only twenty years old, seemed almost pre-historic (22-24).

Curious elements of the structural system are the large free-standing columns at each gable end supporting the ridge. Typical of the Ise shinmei-style, they may derive from the earliest days when the walls were non-bearing and columns supported the weight of the central ridge beam. Or they may be a remnant of a time when the entrance was at the gable end and the post supported a wide overhang over the doorway.

Other evocative symbols of the distinctive Ise Style which reflect ancient construction methods are the *chigi*, crossed members at the roof ends, natural extensions of the tied rafters of early roof construction, and the cigar-shaped *katsuogi* along the ridge—remnants of early weights for securing the thatched ridge (25, 30).

Combining the Shinto veneration of nature with the Japanese preference for simple, straight lines, Ise firmly established the integration of natural and geometric form —which eventually became the hallmark of all Japanese architecture.

The original Japanese roof constructed by legendary deities, as sketched by Hokusai.

ANCIENT ARCHETYPES

For more than thirteen centuries ancient Japanese house and storehouse forms have been carefully preserved as the sanctuaries of Ise Shrine. The sacred center of the Shinto religion, Ise has been a mecca for every Japanese for over a thousand years. Its archaic forms, therefore, are an innate part of Japanese sensibilities (17-29, 40).

No study of Japanese architecture can ignore Ise's influence—the preference for wood as the primary building material, the veneration of natural materials and plain surfaces, the integration of nature and architecture, and the celebration of expressive structure. Ise was, in short, the archetype for Japanese architecture.

Ise consists of two major shrine complexes, the inner shrine or Naiku, and the outer shrine or Geku. Most of the photographs show the Naiku. Built on white gravel clearings deep in an ancient forest, the shrines have been rebuilt nearly every twenty years since 690 AD. After dedication of a new shrine complex on the adjacent site, the old shrine is dismantled. Some of these photographs show a brief time during the 1953-54 reconstruction when both the old and new shrines existed side by side.

Ise may be the most spectacular, but it is not the only place where ancient building forms have been reverently preserved—other shrines and temple treasure houses, some of great antiquity, also reveal the foundations upon which the folkhouse tradition rests (30-39).

The central shrine buildings of the twenty-year old Naiku shrine at Ise (17), and a forest path leading to the shrine (18). The new shrine on the eastern site after clearance of the adjacent site (19).

Steps mount up to the massive Ise style torii arch which marks the entrance to hallowed ground (20). Beyond the torii is the inner gate (21) showing the mixture of elegant and elemental structural techniques used at Ise.

The inner compound with the main shrine and its flanking treasure houses—replicas of ancient store houses and perhaps early palaces. The raised forms are not only a practical means of protecting the contents, but add a monumentality and segregation appropriate to this celestial residence, while the huge columns buried in the ground express Ise's innate connectedness with the sacred earth.

The rather discordant structure over and in front of the stairs is a later addition providing shelter for the frequent ceremonies. Beyond is the newly dedicated shrine on the adjacent, slightly higher site.

Even though the essential form and structure of ancient times is faithfully preserved, a primitive simplicity perished as parts and details were refined during the many rebuildings. The crossed chigi are still extensions of the rafters (25), but the cigar shaped katsuogi have long since lost their original purpose of weighting down the thatch ridge and are now merely evocative symbols of Shinto shrines. Other subtle refinements of form, such as the curve of the roof are particularly evident in the crisp, new Naiku shrine soon after dedication (24).

On the opposite page, dismantling of a smaller shrine reveals the complexities that have evolved since the crossed chigi were originally two rafter poles lashed together with straw rope. At the lower left is the meticulous member (here upside down) devised to hold aloft the now purely decorative katsuogi. On the right, are the delicate parts underlying what was originally a simple thatched roof (25).

In the nearby forest, a miniature version of the main shrine (26, 27 top) shows the curious free standing post supporting the central ridge pole. It is curious because, practically speaking, it is structurally superfluous, yet it has always been an element of this style. Its use may stem from a time when the entrance was at the gable end and the column supported the extended ridge pole and roof sheltering the door. The eight small rods now projecting from the outer rafter originally helped secure the ridge covering.

The beauty of this simple, direct structural system is also evident in the other buildings on the shrine grounds—a washing and purification shelter (27 bottom), the interior of a similar shelter (28), and a ceremonial pavilion with a board on board roof (29).

While the Ise shrines are the main guardians of the pure Japanese tradition, shrines in a similar style were built throughout the islands. At Shinmei-gu, near Matsumoto (30-33), the dramatic setting and the simple, straight lines of the roof evoke a sense of great antiquity. Closer examination, however, reveals Buddhist style corruptions in the small brackets atop each corner column (33) and in the collar beam just below—details not found at Ise.

This small mountain valley marked by giant trees is a perfect setting for the nature worshipping Shinto tradition— and the subordination of the architecture to its site is probably more typical of ancient shrines than the rarified precincts of Ise.

Izumo, the second most important Shinto shrine and some believe the most ancient site, has been rebuilt several times. Tradition has it that the original was considerably larger than the present building, which has also lost the straight lines of the roof and integral chigi and katsuogi—both now purely symbolic and wired precariously in place (34,35).

One other ancient building technique has been preserved in grand scale in the Shosoin at Nara. A true treasure house containing the Emperor Shomu's regalia from the eighth century, the Shosoin's massiveness makes it relatively fireproof and its elevation above the damp ground keeps it dry.

Similar interlocking log walls, though at a much reduced scale and more highly refined, are also still used on some smaller treasure houses at Ise— remnants of its use in the original shrine. Even though this is wall bearing construction—there are no exterior posts supporting the roof—the structural expression is still clear. The massive posts and extended floor beams not only provide a powerful rhythm but clearly express the work of support and contrast handsomely with the horizontal patterns of the wall above (36-39).

Walls surrounding the main shrines at the Naiku, Ise (17-29, 40).

FOLKHOUSES—MINKA

Late one cold and rainy fall day many years ago, I was driving through northern Japan near Tsuruoka exploring the countryside for old farm houses and villages. On an impulse I followed a long tortuous trail—it could hardly be called a road—ending in a remote mountain valley.

I halted in the middle of a dozen small farm houses surrounded by terraced fields. The farmers and their wives, just returning from the fields in their old-fashioned straw rain coats and hats and surprised by my appearance in their village, inquired if I had lost my way. One man, taking pity, invited me home for dinner and we walked in silence to the far edge of the village. His house, typical of this mountain village, was small and old but sturdily built of wood plank walls and thick posts supporting a ragged thatch roof sagging from the rain.

My host slid open the large wooden door and announced our arrival. We entered a room that seemed to occupy nearly half the house and opened to the beams and thatch above. The floor was hard packed earth with a raised section of smooth wood and straw mats.

The foreign guest caused some whispered instructions and embarrassed giggles from the three children. His wife, apologizing for such an old house and the meagerness of the meal, added an extra place around the charcoal pit in the wooden floor.

We sat down on the edge of the raised floor, removed our shoes and then everyone gathered around the hearth over which hung an iron kettle of boiling tea water. In the dim light from the open door I could make out the roof beams and the drying rack above the hearth, all encrusted with generations of black soot. Through the doorway mist-shrouded fields and mountains disappeared with the fading daylight. Inside the only palpable warmth came from the tea cup clutched in my hands. In a few moments it was dark and we ate silently, a ring of faces lit by the glow of the charcoal embers.

No evidence of the twentieth century intruded—no radio or television, no electric light, no sound except the rain and the boiling kettle. In the space of a few minutes I had stepped from my automobile back several centuries in time. Here in this valley, bypassed by the modern world, I had stumbled upon a last remnant of life that had endured in the Japanese countryside for more than a thousand years.

JAPANESE FOLKHOUSES describes a legacy of that way of life—a folk architecture of uncommon power and beauty with a remarkable relevance to modern architecture. It is a relevance not of specific forms but of abstract concepts of space, structure and form that underlie not only folk houses but all Japanese architecture.

Folk style buildings in Japan are called 'minka' (miing-ka). Not all minka are houses of the common folk, and not all common houses are minka. The term minka denotes buildings with certain distinct characteristics: rugged structure, simple rustic details, and bold form. Whether found in farm houses, mansions of wealthy landowners or merchants, urban residences or inns, these characteristics distinguish minka from both ordinary houses and from the refined *sukiya* and *shoin* styles of upper class homes and palaces—even though the materials and construction techniques are similar.

Enduring Minka

Thousands of authentic minka survived into the mid-twentieth century because the entrenched feudal system and poverty of the countryside discouraged change. Radical shifts in Japanese society and the unprecedented prosperity of the last 20 years have revolutionized rural life, and minka were among the first casualties.

The reasons are not difficult to ascertain. The old farm houses, products of a different age, do not adapt well to modern conveniences. They tend to be dark, difficult to heat, and, perhaps even more devastating, they symbolize the yoke of grinding poverty and feudal oppression of the not-so-distant past. Living in these old houses today requires sacrifice and considerable extra expense for maintenance. Younger family members, when they remain in the village, often refuse to cope with the inconveniences. The owner will likely either radically make over the old house or tear it down to build a modern house equipped with all the latest appliances. He will be egged on by his newly 'liberated' wife who, tired of coping with a kitchen barely out of the middle ages, demands a sink with running water, gas stove, washing machine and refrigerator. That most pervasive force of all, television, has spread the foreign and urban lifestyles that make the old homestead seem drab and outdated. Although the new houses seldom have any style, they do confer status—just as the former minka were important status symbols in their time. Certainly remaining in the old homestead prompts nothing but sympathy. Minka survive, therefore, mainly by inertia and sporadic efforts at restoration—or as weekend retreats for city dwellers.

Historical Background

Few minka more than two or three hundred years old exist today. Wooden temples dating back a thousand years still stand, but no houses. Timber construction and thatch roofs are too susceptible to decay and especially to fire. Typhoons and earthquakes also have taken their toll, sometimes from the devastating fires in their wake. And throughout history battles between warring clans frequently destroyed whole villages or ravaged the countryside. Not only catastrophic losses, but the evolving styles of minka and the changing fortunes of their owners caused periodic replacement of deteriorating or inadequate older structures.

Existing minka, therefore, represent the final stage of development—the product of the peaceful and stable last two and a half centuries during which the authority of the central government and improved communications blurred many old regional differences.

Our knowledge of the earliest houses comes from archaeological reconstructions, early written descriptions or drawings, and from the archetypes at Ise. Existing minka, although vastly increased in size and complexity over those ancient types and incorporating many new refinements such as mat flooring (*tatami*) and movable room partitions (*fusuma*), still retain many of the essential traits of the primeval forms.

Rice Culture

Since recorded history Japanese wealth

Harvesting activities—cutting, threshing, and winnowing—in a Hokusai woodcut.

and power depended on rice and the land on which it grew. Most farmers were tenants. Whole villages often served a few landowners who collected forty to sixty per cent of the rice crop from their tenants and were in turn taxed by the next higher authority. In this rigid social order each class had its place and was expected to keep it—and the farmer's place was producing the rice upon which the whole system relied.

Other sources of income in the countryside included lumber, charcoal, and silk and handicrafts such as pottery which were traded to the townspeople. But trade was held in low regard. Only in the nineteenth and twentieth centuries did the merchants acquire the money and power to gain respect, though the taint of being 'in trade' hung on until after the last war. Many of the finest urban minka were built by these prosperous merchants (145-153).

Village Life

'Japan's mini-farmers never have been independent entrepreneurs in the way that the capitalist farmers of the West, living in isolated houses surrounded by their own broad acres, used to be. The settlement pattern in clustered villages, the crucial dependence on rice agriculture with its shared irrigation systems involving collective village control of water and who could use it, the juxtaposition around each village of tiny plots whose use or misuse could affect the plots of a neighbor, the intricate patterns of labour exchange between households—all created a pattern in which the use of land was a right and a responsibility shared between the individual household and the village community.'[2]

The regimentation of Japanese society and

the importance of cooperation in rice agriculture made the village the natural center of rural life. Initially villages were fiefdoms under the absolute control of a *daimyo* who issued regulations, settled conflicts, and supervised the upkeep of roads, bridges and irrigation works. Later the villages governed themselves through headmen appointed from the wealthy landowners. Families tenant farmed the same scattered fields for generations. A few owned small plots in addition. Although in principle land could not be sold or mortgaged '(how dare they behave as if they owned the land which has merely been allotted to them to perform their heaven-sent duty to cultivate!)'[3], the regulations were impossible to enforce and land constantly changed hands according to the altered fortunes of its owners.

Intensive use was made of the land. Rice paddies, diked and terraced to hold water, were painstakingly constructed in all possible locations. Tiny leftover plots were planted with vegetables and other crops, and the hills around the village were maintained as wood-lots or bamboo groves. Houses, whenever possible, were built on land unusable for growing—often along the base of hills at the edge of the fields.

Life in the villages was hard and, in times of poor harvest, many families verged on starvation. The seasons set the pattern of rural life—the toil and tedium broken only by frequent festivals celebrating spring planting, fall harvest or religious events (72-73, 108, 166-167).

Integration with Environment

Characteristic of folk architecture in general is the thorough integration into the life and environment of its time and place. Japan is no exception. So completely enmeshed are Japanese customs and domestic architecture that the cause and effect relationship between them is difficult, if not impossible, to determine.

An example of this close relationship between built form and social custom is the attitude toward privacy and the extreme openness in the Japanese home. From prehistoric times Japanese houses have been essentially one large room. The post and beam support eliminated the need for solid interior walls. Instead interiors, minimally divided by thin movable walls, provided little personal privacy—the source for many amusing and ribald folk tales. Whether such physical arrangements were the cause or the effect, the Japanese contend with the lack of privacy by a remarkable ability (or pretense) to not see what they do not wish to see, and not hear what they do not wish to hear. Many cultures have had to cope with one-room houses, but none so artfully as the Japanese.

A similar cause-effect relationship, already discussed above, is the connection between the lack of physical comfort in the typical Japanese house and idealization of stoicism in Japanese life.

The architectural consequences of this unity between folk life and folkhouse was to make the house an important symbol in Japanese society. In many vernacular environments, such as the white cubist towns of the Mediterranean, houses were anonymous clusters of undifferentiated shapes with minimal opportunity for individual expression or symbolic display. In Japan, however, free-standing houses surrounded by their own compounds or gardens offered greater freedom to express the singular needs and desires of each family.

Strangely, Italian villages, full of individualists, were uniform in character, although Japanese villages, where individualism was discouraged, were full of dramatic, expressive houses. Perhaps one compensated for the other.

Folkhouses as Symbols

The strict class distinctions in Japan applied to clothing, manners, speech and to houses—perhaps the most prominent symbol of ones status. In an effort to maintain class distinctions dwellings were the subject of frequent official edicts which placed limits on the maximum size of houses, on the height and style of roof and length of its ridge, on the use of certain materials, decoration and styles, and even on the finishes and room types allowed for each class. The real purpose, of course, was to prevent peasants from building houses obviously beyond their station in life. It was a futile effort. For instance, peasants in some districts were limited to small beam sizes (incidentally preserving scarcer large timbers for the upper classes) but they devised ingenious structural techniques to circumvent restrictions and build roofs as large and grand as desired.

Teiji Itoh relates two specific attempts at regulation: 'In 1658 the Makino family, castle lords in Niigata, issued certain regulations concerning building styles for all their retainers. These regulations stipulated that the houses of all people ranging from a managerial level and the level of a samurai (warrior)... down to the level of lower-class warriors must all have roofs of miscanthus and had no need for the chumon style with its L-shaped structure for outbuildings for storage.' And again: 'In 1740, the Murayama clan issued an edict to the effect that all farmers' houses must use posts of no greater length than (4.24 meters), that the roof must be thatched with straw, that ceilings must be of the slatted type, that the floors must be of pounded earth, and finally that neither the L-shaped plan nor outbuildings were necessary.'[4]

Such jealous guarding of architectural status symbols appears to be age-old—an early Japanese chronicle tells of a lord who, coming upon a farmer's house that dared to ape the details of the lord's own roof, was barely restrained from mayhem.[5]

Further evidence of the integration of architectural form into the rigid class system was the requirement, at least in the homes of landowners, for separate entrances for each class—an elegant, formal entrance for samurai, a middle entrance for 'middle' classes, and the utilitarian doorway used by family, peasants, and tradesmen. No one would consider using an entrance above (or below) his class (54, 111, 185).

Regional Variations

Japanese geography and the conflicts of a feudal society forced long periods of regional isolation, often for generations. Isolation intensified regional variations in the minka tradition because either time stood still in some forgotten backwater or, more often, because local innovations matured into distinctive styles.

Innovation, normally resisted in traditional societies, represents a threat to the established order. Technical and aesthetic innovation must have been especially difficult within Japan's inflexible social and guild structure. Despite the difficulties the building tradition was constantly evolving—sometimes forced by need or sometimes by an exceptionally creative individual whose fresh ideas, repeated often enough, eventually made a permanent change. Such was the case with Takayama's renowned carpenter-builder, Kawajiri Jisuke, who built some astonishing houses for the merchants of his town in the late nineteenth century—houses wonderfully full of light in radical departure from the typical gloomy minka interiors (145-153).

The Takayama houses benefited from the recent introduction of glass, just as much earlier the availability of paper opened dwelling interiors by using light-transmitting *shoji* panels on the exterior. Not only new techniques but the special need of a merchant, the abundance of a particular material or skill, or the perverse regulations of local authorities all challenged the established way of building.

Such a mixture of local needs and talent caused a radical departure from ordinary house types in the remote valleys of Shirakawa (154-159). A combination of large communal families with 25 to 30 members, limited building space in narrow mountain valleys, silk culture, and the talents of the nearby carpenters of Takayama resulted in a highly specialized form of Japanese folkhouse: large, steep roofs covering two floors of living space and two or three attic floors housing trays of silk worms warmed by the inhabitants and animals below. It was an efficient solution to these particular needs but, when the silk culture died out and the family system declined, the huge houses became a burden to those left behind—the dinosaurs of a bygone age.

Materials and Construction

Japanese folkhouses were constructed of natural materials readily at hand—wood, clay, straw, bamboo, and reeds. Despite two thousand years of working within this limited palette, the technical and visual potentials seem inexhaustible—given the infinite variations of natural materials and the ingenuity of superb craftsmen.

The basic materials and the construction methods common to minka and to all traditional Japanese architecture have changed little over the centuries. The primary frame consists of columns resting on large stones set in the ground supporting a complex network of roof beams, which is overlaid by a lattice of bamboo or twigs and covered with thatch, wood shingles, or clay tile (114, 157).

Walls between the posts are made of thick wood planks or layers of clay applied to both sides of bamboo lath (103). Small windows, covered by rhythmic wood grills echoing the structural pattern, pierce these walls (134-135). Large sections of the exterior are covered only by one or more layers of sliding panels. The inner panels, shoji, light wood frames covered with translucent paper, provide light and privacy for the interior. These shoji are easily opened or removed (173). For more security solid wood panels, *amado*, stored in pockets to the side, cover the shoji at night.

(continued on page 138)

SAIGOKU
KYOTO AND THE WEST

Kyoto, the capital of Japan during most of its history, and nearby Nara, the previous capital, are at the center of traditional Japanese culture. In the region from Kyoto south and west along the Inland Sea to the great Shrine of Izumo the essence of Japanese architectural tradition evolved.

In this crucible of land and sea the legacies of aboriginal pit houses, south Asian pole houses, and sophisticated ideas from mainland China blended into a distinctive folk style adapted to the mild climate of the region.

As the photographs convey, the Japanese folkhouse, despite the elemental nature of its materials, is more than a primitive shelter of sticks and straw. Like many things in Japanese art and life, the obvious practicality of the folkhouse only partly obscures an obsession with form, style, and symbol.

The myriad variations in roofs and other details belie the underlying unity of the region's indigenous forms—a unity which reflects the similarity of climate and geography and the dominance of Kyoto's political and cultural influence.

The soaring thatch roof and highly decorated ridge on this small farmhouse testify to the importance of the roof in Japanese folkhouses both as status symbol and as a source of pride for its builders (49).

At Nose, in the Tamba district west of Kyoto, is perhaps the oldest folkhouse or minka in the book—claimed by the present owners to be more than four hundred years old. It certainly looks very old with its slightly rough-hewn quality, brown mud walls and massive roofs. The entrance gate on the left and the storehouse (kura) on the right enclose the courtyard in front of the main house beyond (50-51).

Isolated localities evolved their own local styles and techniques—especially in roof forms. For example, in mountain regions near Kyoto a distinctive gable roof style developed. Smoke is exhausted through a low cupola and end walls are boldly half-timbered. Wood shingles, the favorite roofing of ancient times, are still used—both shingles and thatch long since banned in towns because of the fire hazard. Keihoku (52).

A cluster of kura and thatched houses in what may be either a tiny village or the domain of an extended family. Yoshitomi near Kameoka (53).

At Ibogawa on the Inland Sea is the Nagatomi house (54-61), an especially fine example of an early 1800's landowner's house. It was extensively restored in 1963 and the forecourt turned into a sophisticated garden. The owners' elevated social position is evident from the three entrances—a formal entrance for samurai guests at the left, the regular guest entrance in the middle, and the family or tradesmen's entrance to the workroom (58) at the right.

The original roof was also tile, because by this time it had become the prestigious roofing for wealthier households.

Spacious tatami-floored living and guest rooms open onto a sheltered veranda, or engawa, overlooking a walled garden (55).

The earth-floored room just inside the door (58) was a large space for work and storage, and where the landowner conducted business with his tenants. Overhead the open roof beams let in some light and allowed smoke from the constant cooking fires to escape. The roof structure over this space was always exposed in traditional houses so beams were selected as much for their impressive size as for structural necessity (59).

The adjacent mat-floored rooms (56-57) for daily family living centered around the kitchen at the rear (60).

While this is certainly not an average folkhouse in size and elegance, it is built in the same general plan and with the same spaces one finds in even the poorest house.

Ovens for preparing large batches of rice are the main elements in the kitchen. Kept stoked with twigs and straw, the large masonry and plaster mass retained the heat and provided one source of warmth in winter in an otherwise unheated house—though the heat quickly dissipated up through the roof along with the smoke (60).

The sophistication of the owners is apparent in the extensive tatami rooms complete with art works. Because the rooms were used for eating, sleeping, and entertaining, and, because it was customary to sit on the floor, there were almost no permanent furnishings. The light-weight sliding panels between the rooms (fusuma), while giving little privacy, allowed great flexibility in combining spaces and were completely removable for large groups (61).

Compact villages west of Nara, one of the most densely built-up regions, form striking patterns of alternating roof planes and are the closest in feeling that the Japanese village comes to Mediterranean hilltowns. Near Ikoma (62 top, bottom), and Omiya near Sakurai (63).

Like most villages, those between Nara and Kyoto began as clusters of a few farmhouses. As the village grew, the houses became tightly packed between the precious rice fields leaving only narrow streets for access. The only public spaces for village gatherings were around the local Shinto shrine or Buddhist temple. Most houses, however, maintained private courtyards behind the walls. Villages in the Ikoma area (64,65).

In the Nara area the use of tile is more common, perhaps the influence of the time when Nara was the Imperial capital and a Chinese tile roof was a mark of distinction. A roof style unique to Nara has plastered end walls capped with tile and a tile-ridged thatch roof between. This form may have developed partly to lessen the spread of fire in the crowded villages of the Nara plain and partly for its decorative effect (66).

Though Tamba is famous for its huge thatched roofs, some localities mixed spectacular tile roofs with thatch as here in the village of Akakuma near Yagi (67 top & bottom).

In contrast with the usual pattern of houses clustered in villages requiring long treks to the fields, around Wachi west of Kyoto, houses stand amidst the fields (68, 70, 71).

Bamboo, as beautiful as it is useful, has been an important material in Japanese rural life. Mountain streams were piped to the paddies through its hollow stems. It was split for gutters at roof's edge or into strips woven as lath in plastered walls. It also was used structurally as the final layer under the thatch, decoratively in fences, and not least, bamboo shoots were eaten as a delicacy in early spring (69).

Since ancient times rice has been the staple food of the Japanese and central to Japanese life. It was the primary source of wealth and also the principal means of exchange for centuries. Taxes were paid in koku (bushels) of rice and a man's wealth and social position were determined by the koku of rice he commanded.

The planting and harvesting of the rice were special times. Groups of women, even today, dressed in traditional kasuri for the occasion, work long hours bent over in muddy paddies (73 top right).

Men preparing rice paddies on the slopes overlooking Lake Biwa (72); a pause in plowing (73 top left); working in the old straw rain coats and hats (middle left); again at harvest time, the women wear the handwoven indigo and white costume while cutting the grain by hand and hanging it on racks to dry (bottom left and right).

Where there is rice there must be plentiful water and in Japan water seems to be everywhere in streams, lakes, and paddies. Shijo village (74) near Nara and a house near Lake Biwa with a newly planted paddy at its doorstep (75).

In the high country the paddies were carved out of the hillsides and the houses fitted among them. A village between Keihoku and Yagi (76-77).

Repairing flood damaged terraces on the northern outskirts of Kyoto (78). Sasayama (79).

In a land as mountainous as Japan the lowlands were too valuable to use for building so in the older parts of Japan the villages tend to be built just above the valley floor. In some cases (80) they stretch out along a road at the base of the hills creating a rhythmic geometric separation between the wild hills above and meticulous fields below. Ikoma (80), and between Uji and Nara (81).

Rice fields crowd up to the main gate of a large farmhouse in Hozu-cho (82, 83).

The villages of Tamba, a land crisscrossed by mountains and fertile valleys, are often dispersed—as in this valley between Wachi and Miyama where houses line both sides (84, 85).

Many farmhouses are surrounded by walls with an impressive gatehouse—not so much for security as for prestige and privacy. Inside, the yard is often used to dry crops in the sun. Between Ibaraki and Kameoka (86), near Lake Biwa (87).

In contrast with many other parts of the world where the walls or mass of a house first catches our attention, in Japan it is the roof. Such huge roofs were not only practical in this rainy climate but, perhaps even more to the point, were a source of pride both for the owner and the builder. These great forms were landmarks in the landscape—and since ancient times symbols of authority and wealth. In the rigidly controlled society that was old Japan a roof was one of the few ways to proclaim wealth or individuality, but even its construction was often proscribed so that one could not pretend to rise above his station.

Traveling around Japan, it is fascinating to follow the changes in roof style from region to region. While there is a certain uniformity across wide areas, local variations on a theme, often the inspiration of some local craftsman, are endless. The next few pages are a cross section of roof styles and details from southwestern Japan, the cradle of Japanese architecture (88-101).

Major replacement of thatch roofs is required every ten to 15 years, when the ridge and gable ends are also renewed. Rice straw is sometimes used, but the best thatching materials are certain reeds and other types of straw. Special crews traveled a region, working on each roof for several days. But such skills are scarce in modern Japan, so the thatch roof has become something of a burden and is rapidly disappearing. Since the thatch is usually full of vermin, many families are only too willing to replace it with tile or metal. Between Keihoku and Hanase (88).
Tamba, between Ichijima and Miwa (89).

Roof style used near Hiroshima (90).
Special brackets support extra wide overhangs in this house on semi-tropical Kyushu, protecting the sliding paper walls (shoji), the wood-floored engawa, and the space beyond from the typhoon rains and from the summer sun (91).

Farm house groups, in which the roof planes make powerful compositions of form and texture dominated by the great thatch roof of the main house. Tamba (92) and Miyama-cho (93).

Farm houses along a narrow valley in Miyama-cho. The elevated wood roof of the fireproof storage house or kura protects the plastered top from the elements (94-95).

One of the few remaining places with a large number of simple thatched houses, this village in Miyama-cho retains the authentic character of ancient times (96-97). When possible, as here, the houses were built with the long entrance side to the south for maximum winter sun. To prevent blocking of the sun by a neighbor's roof the houses are dispersed up the slope near the village shrine set in the grove of trees at the center.

Details of roofs from southwestern Japan (98-101). Tojo, (98 top), Sadamitsu, Shikoku (98 bottom), and Tamba (99). The gable ends provide a place to vent the interior, to decorate, or to display the family crest.

Interim roof repairs involve inserting new thatch in the old (dark layers) and then trimming the whole (100 top).

Tiny mountain village on the road from Himeiji to Tohara (100 bottom), a minka on the main road of a Tamba village (101 top), and a country house in Miyama-cho (101 bottom).

Beneath the roofs are large, open, living spaces—such as this typical interior at Miyama-cho. The earth-floored room just inside the door contains the kitchen with its rice oven at the left. The family gathers for meals and wintertime warmth around a small cooking pit in the raised wooden floor at the center (covered over for the summer). Years of wiping away the soot from the open fires have polished the walls and posts as far as the wife could reach. Wooden sliding doors divide the spaces. The tatami-matted rooms in the rear are for sleeping or entertaining and since no shoes are worn on the raised floors they stay clean and smooth. The spartan look of the interior is characteristic (102).

Ante-room of a large house in Horyu-ji village near Nara, where the work room is closed off by sliding grills and guests enter directly to the living space at the right (103 top).

Wearing kasuri, this woman, just returned from the fields, stands next to a damaged section of her barn that reveals the construction methods of all traditional Japanese architecture—whether folkhouse or palace. A heavy wood frame of columns and beams is left exposed while thin horizontal ties are overlaid with split bamboo lath, then covered with plaster (103 bottom).

A particularly fine use of tile and thatch roofs where the tile portion extends the main living space. The curved logs of the ridge give a dramatic lift to the roof (104).

This storehouse near Takao, Kyoto, combines a beautifully fitted and curved stone base with wood siding (protecting the most exposed part of the wall) and molded plaster. Despite the kura's massive appearance, beneath the fireproofing layers of plaster is the usual wood frame (105).

Tamba houses behind stone walls built, not for any defensive purpose, but as retaining walls for high ground at the edge of the fields. Hojo-ji (106), Sasayama (107).

Every village has its Shinto shrine and annual shrine festival when the deity is paraded on a palanquin by young men of the village. The torii arch marks the entrance to the shrine precincts, often a small stand of woods near the village. Tamba (108 both) and Shitsukawa, Tamba (109).

Yoshimura-tei, between Nara and Osaka, is the restored 17th century house of a wealthy landowner. Now a national treasure, it is certainly one of the finest houses in Japan. It combines the simplicity of ordinary farmhouse construction and layout with the sophisticated interiors and gardens of the upper classes. The south facade, just inside the gate house, has the required separate entrances—on the far left, the formal samurai entrance, in the center an entrance for ordinary guests, and on the right, the doorway for daily use of the family, farmworkers, tradesmen and the oxen whose enclosed stall was inside (110-121).

At Yoshimura-tei the asymmetrical rhythms and elegant proportions of the structural pattern integrate a variety of functional elements. This shows how richly gratifying form may derive from simple, utilitarian means—a characteristic of vernacular building and a principle source of expression in all Japanese architecture.

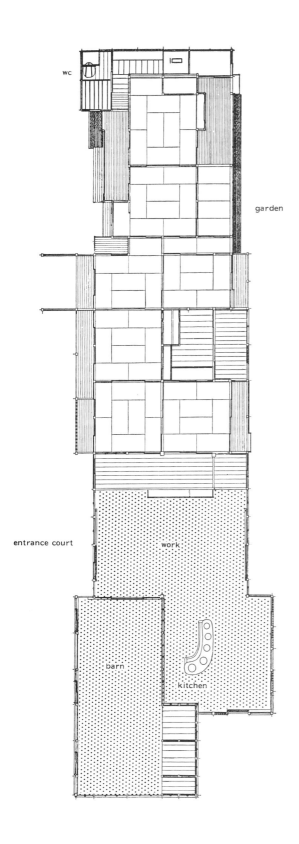
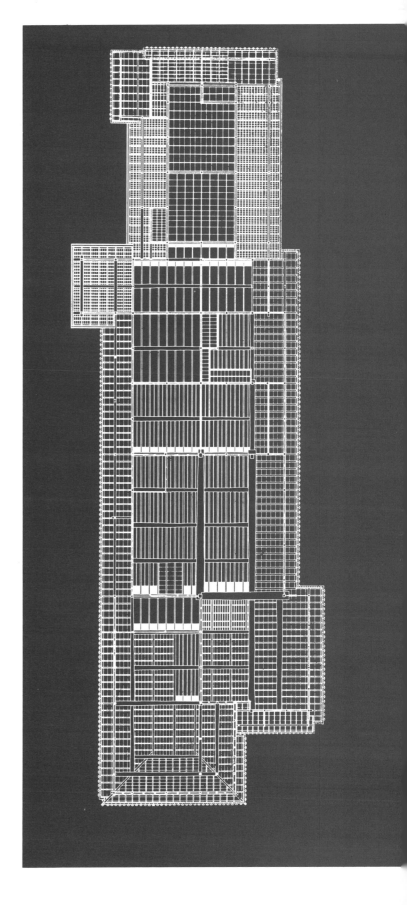

Posts and walls are set on stones to protect them from the damp ground (114).

Floor plan and reflected structural plan, Yoshimura-tei (115).

The earth-floored room with kitchen and work space. In this busy place the maids (who slept in the suspended room at the rear) prepared meals, washed and dried foods. Here also, tools were repaired, peasants brought the landlord's share of rice to be measured and stored, and the animals came and went from the small barn at the left.

Toilets, when attached, were kept remote since the night soil periodically was removed to spread on the fields. Engawa leading to the toilet (118).

Fusuma, opaque light-weight sliding panels divide the interior rooms. At the exterior, translucent paper shoji doors let in daylight and are covered at night by amado—solid wooden panels.

The openness and the movable walls meant the interiors were impossible to heat in the winter. No effort was made to warm whole rooms; instead, families gathered around the cooking pit or charcoal hibachi. Sometimes a charcoal burning kotatsu, recessed in the floor beneath a table and covered with a quilt, warmed arms and legs, while the deadly fumes were carried away on the drafts.

Faced with extremes of climate, the Japanese stoically endured drafty, unheated interiors during the cold winter in order to throw open the house to the breezes in the hot humid summer.

Yoshimura-tei facade (120-121).

Urban minka faced different conditions. The lack of building space and need for more fireproof construction resulted in long narrow row houses, tile roofs, and plastered dividing walls. Still the towns were frequently ravaged by fires that burned out whole sections in a single night. With minimum frontage on the street, use of wood grills satisfied the occupants' need for light and air and yet provided security and privacy to the interior. Variations on this urban theme are found throughout Japan but the finest examples are in the older parts of Kyoto.

Town house in Takehara, near Hiroshima, where nearly the whole street facade is covered with grills in various patterns whose rhythms are echoed in the tile roof (122).

In Kurashiki, known for houses built in the kura or storehouse style (124-125), this street elevation incorporates an unusual amount of woodwork (123).

Typical Kyoto minka from the Gion, center of high class Geisha life (126-127).

A combination of infinitely varied grilled windows and fences (to fend off passersby) results in lively but unified rhythms both in the facades of individual houses and of whole streets. Nishijin, Kyoto (128, 129).

The Pontecho (130) and Gion (131), Geisha quarters of Kyoto, maintain the character of old Japanese cities with their narrow streets lined by small wooden houses.

The old styles were preserved as part of the atmosphere in Kyoto's Shimabara, a former red light district. In the entrance court to the Sumiya Inn the open doorway is covered by a cloth noren decorated with the inn's crest (132). Several layers of surprisingly open grillwork cover the shoji panels on its two-story exterior (133).

Window styles from city and country folkhouses, all with some variation of the ubiquitous grill—Kyoto (134), Tamba (135), and Shimabara (136).

Inner walls of minka, originally solid wood or sliding wood panels, are now more often made of fusuma, lightweight paper-covered solid doors (61, 102, 119). Free-standing screens and grills are also used for partial privacy (103, 145).

Floors

The hard earth floor of Japanese prehistoric dwellings is still a part of every minka because it is economical and practical for work areas and kitchens (58, 60). The floors in the living and sleeping rooms, raised 18 to 24 inches off the damp ground, were originally surfaced with smooth wood planks, but are now mainly tatami—resilient straw mats covered in woven reeds (102, 116, 190).

The tatami mats, along with sliding interior walls and paper-covered shoji, which replaced the simpler and more rugged floors and walls of earlier minka, were borrowed from the elegant homes of the upper-classes. As use of the three foot by six foot tatami spread, it became the basic design module. Rooms were specified by their mat size. This standardization of the tatami and of column placement encouraged prefabrication not only of tatami but of modular shoji, fusuma panels, and wooden structural elements, though the standards varied slightly from region to region.

The Roof

The history of the Japanese minka is a history of roofs gradually increasing in size and grandeur, encompassing more and more space within their wide overhangs—including space beyond even the walls themselves. From the simple teepee-like roofs of ancient pit houses, to the steep roofs of Shirakawa, to the elaborate roofs of northern Honshu, roofs symbolized shelter and stood as evocative landmarks amidst the fields and hills (68, 70).

One pleasure in traveling through the Japanese countryside is discovering new roof styles and the myriad variations on the local theme. Proud of this most prominent part of their houses families spared little effort or expense on the roof and its ridge decoration. As a distinctive local trait, roof styles did not

Hokusai's woodcut of a New Year's celebration around the hearth in the earth-floored room. In the background are the kitchen ovens and two large straw bales of rice.

spread much beyond their original territory.

Although there are hundreds of local styles roofs took three basic shapes: simple gable (52), hipped (161), and the most common, a hipped-gable style called *iri-moya* (49). The latter, a form depicted on prehistoric bronzes, included small gable openings to exhaust smoke from the interior. All the styles frequently added lean-to roofs, tucked below the main eaves, covering verandas and extensions to the main building. Broad eaves, extending several meters beyond the house proper, shaded interiors from the sun and kept out wind-driven rain (91). The photographs graphically illustrate both the hovering dominance of these roofs and their visual importance in unifying the diverse elements and patterns beneath.

Thatch made of easily gathered reeds and straw was the most common roofing. Thatch required repair or replacement every 10 to 20 years though the rare miscanthus reed lasted a lifetime. The simpler roofs and interim repairs were made by the farmers themselves. Major re-roofings or elaborate roofs required specialized crews that roamed the district (88, 100).

Another favorite material was thin wood shingles, still used in some mountain districts (sometimes weighted by stones) (52, 168). Both thatch and shingles were long ago outlawed in the towns because of fire. They were replaced by tile, introduced from China in the 7th century. A serious problem with early tile roofs, laid in a bed of plaster, was the added weight which increased the threat of collapse in typhoons and earthquakes. Tile may lack the rustic charm of thatch or shingles, but it has its own beauty and is more adaptable to varied roof shapes.

Thatch, on the other hand, works best in simple, symmetrical roof shapes typical of minka. Irregular requirements of the plan are accommodated by moving walls in and out within the limits of the eaves, or by small extensions of subsidiary roofs. The regularity and symmetry of these roofs were based on

Owing to the former somewhat isolated life of the different provinces, the style of building in Japan varies considerably; and this is more particularly marked in the design of the roof and ridge. Though the Japanese are conservative in many things concerning the house, it is worthy of note that changes have taken place in the house architecture within two hundred and fifty years; at all events, houses of the olden times have much heavier beams in their frame and wider planks in their structure, than have the houses of more recent times. A probable reason is that wood was much cheaper in past times; or it is possible that experience has taught them that sufficiently strong houses can be made with lighter material.

The Japanese dwellings are always of wood, usually of one story and unpainted. Rarely does a house strike one as being specially marked or better looking than its neighbors; more substantial, certainly, some of them are, and yet there is a sameness about them which becomes wearisome. Particularly is this the case with the long, uninteresting row of houses that border a village street; their picturesque roofs alone save them from becoming monotonous. A closer study, however, reveals some marked differences between the country and city houses, as well as between those of different provinces.

The country house, if anything more than a shelter from the elements, is larger and more substantial than the city house, and with its ponderous thatched roof and elaborate ridge is always picturesque. One sees much larger houses in the north,—roofs of grand proportions and amplitude of space beneath, that farther south only occurs under the roofs of temples. We speak now of the houses of the better classes, for the poor farm-laborer and fisherman, as well as their prototypes in the city, possess houses that are little better than shanties, built, as a friend has forcibly expressed it, of 'chips, paper, and straw.' But even these huts, clustered together as they ofentimes are in the larger cities, are palatial in contrast to the shattered and filthy condition of a like class of tenements in many cities of Christian countries.

In the northern part of Japan houses are often seen which possess features suggestive of the picturesque architecture of Switzerland,—the gable ends showing, in their exterior, massive timbers roughly hewn, with all the irregularities of the tree-trunk preserved, the interstices between these beams being filled with clay or plaster. The eaves are widely overhanging, with projecting rafters. Oftentimes delicately-carved wood is seen about the gable ends and projecting balcony. As a still further suggestion of this resemblance, the main roof, if shingled, as well as the the roof that shelters the verandah, is weighted with stones of various sizes to prevent its being blown away by the high winds that often prevail.

Edward S. Morse, Japanese Homes and their Surroundings, 1887.

several standardized structural systems.

Structure and Form

Beneath the great unifying roofs, the powerful rhythms of the boldly expressed structure further ordered the many elements of minka form (111, 112). Because of its high visibility and crucial importance the inhabitants were extremely conscious of the structure. Large scale members were proudly pointed out to visitors. Moreover, in most houses the large central column, *daikoku-bashira*, was a semi-sacred object, decorated with small offerings and polished by the women of the family (102) —perhaps to appease the god of the living tree sacrificed for the column. Similarly at Ise, the trees for use in the rebuilding are blessed by the priests of the shrine before felling and during the decade-long curing.

The massiveness of the roofs made a powerful structure essential; however, an intriguing aspect of Japanese structural development is the conscious avoidance of diagonal bracing—visible or hidden—such as found in European half-timber construction. In a land beset by frequent typhoons and earthquakes such strengthening of wood framing would seem natural. Certainly the plaster walls between the posts contribute little rigidity. Considering the heavy roofs supported on widely spaced columns perched on stones, stability is questionable. Folktales abound with incidents of homes collapsing on their owners during such natural catastrophes.

The lack of diagonals is particularly curious considering vernacular architecture's normal responsiveness to environmental forces—of which typhoons and earthquakes are palpable examples. Three possible explanations have been suggested; ignorance, cleverness, or aesthetics.

Ignorance of the benefits of diagonal bracing is unlikely given the sophistication of Japanese wood construction and the wide use of diagonals in simpler structures (167).

A second explanation which some Japanese claim, unconvincingly, as the true reason for the lack of bracing is that ancient builders very early discovered that flexibility was the key to survival; the inherent resiliency of wood in redundant structural members connected by hundreds of intricate joints absorbed the sudden shocks of earthquakes and wind. There is some validity in the concept but little evidence it was consciously applied.

A more likely explanation is that Japanese architecture, like many Western classical styles, occasionally valued taste and sensitivity over practicality and that sound structural principles were sacrificed to the cause.

The underlying reasons may be as simple and direct as the following: early pit houses with roof framing like giant teepees were naturally resistant to the horizontal forces of winds and earthquake. Once roofs were raised on columns however, the stability against horizontal wracking depended entirely on the rigidity of the columns and walls. In ancient times, columns braced by walls of wooden planks and buried in the earth like those at Ise, were exceptionally strong but soon rotted away. Upon discovering that supporting the columns on stones above ground dramatically reduced deterioration, it was too late for integrating a compensating system of diagonal bracing into traditional aesthetics—the rectilinear pattern had been indelibly fixed.

Moreover, the need and desire for maximum openness and flexibility left few walls in which to install or even hide the diagonals. The result, apparently, was the triumph of taste and tradition over common sense—not the only example in Japanese life or architecture.

Space and Form

Essentially one-story and fundamentally connected to the land, Japanese architecture's bold structural geometry acts to both articulate the horizontal extensions of space and separate the eternal horizontals of roof and earth.

The fluidity of Japanese architectural space beneath the great hovering roofs was a revelation to Western architects of the last century conversant in the defined, compartmented space of classical styles. The idea of space as a continuum in which 'architecture developed with and beyond the structural cage as a series of patterns and planes defining limited portions of continuous space'[6], became Japanese architecture's most important influence in the West.

As the solid, contained forms of Ise shrine and prehistoric one room huts betray, however, these spatial concepts did not spring forth full-blown in ancient times. They paralleled an increasing transparency of the post and beam structure that inspired greater relation between interior spaces and between interior and exterior, culminating in such sophisticated spaces and gardens as the seventeenth century Katsura Palace.

Elegant mansions were the prerogative of the wealthy, leisured classes. Despite attempts to prevent their spread to the undeserving, the more practical aspects of these mansions, such as sliding paper walls and tatami floors eventually permeated every level of Japanese society. Even the simplest farmer's cottage eventually required some semblance of a garden.

The borrowing was not all one way. While the elegant trappings of the leisure class were appropriated for the minka, idealized copies of thatched peasant huts suggesting 'refined poverty' were built in the gardens of the wealthy as a sanctuary for the Zen-inspired tea ceremony.

Although the absence of fixed bearing walls freed spatial planning in minka, practically, the usual row of large interior columns spaced six to twelve feet down the center restricted plans to a few standard arrangements. Only in the earth-floored room where they could interfere with work space were columns avoided by using extra large beams that spanned the entire room (58-59, 116, 170).

Typically houses were divided into two major spaces, the earth-floored work space containing the kitchen, storage, and often a small animal barn; and the raised floor area for family living quarters and guest rooms. In this

An early foreign and somewhat less than objective view of Japanese construction and house design:

...Primeval man may be supposed to have made his first attempt at bridge-building by throwing a log across a stream, and it would not be difficult to prove that the principle of Japanese structures (let them be temples or dwellings for the prince or the peasant) has ascended but little above that embodied in the primitive bridge...
The crude and unscientific disposition of material observable in native structures, notwithstanding their imposing grandeur and elaborateness of detail, in many cases, leads to the conjecture that the constructors were ignorant of the value of braced or trussed framings...They also have ignored the use of diagonal members in their framings, and prefered the rectangular to the triangular division into bays. Some have attributed the absence of diagonal pieces to earthquake considerations on the part of the constructor, but I think it probable that this feature arises more from custom than preconceived intention...
With regard to the strength of purely native structures the conclusions I draw...are that the generality of them are certainly weak, when compared with the quantity of material employed. I also consider them primitive and faulty in the principles of design, and lavishly wasteful in constructive material...

...The question next to be discussed is whether a Japanese house posssesses the properties of comfort and convenience; ...it may be well to treat the matter from a Japanese rather than from a foreign point of view. In doing so, we may presume that the social life of the Japanese has less regard for privacy than ours; and that generally, their faculties of smell and feeling are also less acute. Indeed, reasoning from observation, I am inclined to aver that, in some cases, native sensibility is so indifferent as to bear a range of temperature, which would cause frost-bite at one extreme, and violent fever at the other to a European, with the umost nonchalance and want of bad effect.
Most foreigners, who have experienced them, have expressed their admiration of Japanese dwelling-houses during the summer season, on account of the pleasant coolness they afford; for generally every breeze, however gentle can waft its welcome way through them...The quality which makes them the charm of summer, and the misery of winter, depends upon the simple fact that 'they let in the cold.'...During winter the shelter afforded by the thin porous paper which surrounds the rooms is a mere apology as a screen from the inclemency of the weather. The heat emanating from the brazier of glowing charcoal, is quite local; and only serves to render the pervading chill more tangible by warming one part of the body while the rest is subject to shivering...

In the course of these remarks I have disputed the advantages of the native method of building to some extent; but I have done so upon purely technical grounds and not from national predjudices, and to support this statement I could, if necessary, mention features of Japan of a social and industrial character, which I believe are more worthy of being commended than their buildings are of being condemned.

SOME REMARKS ON CONSTRUCTIONS IN BRICK AND WOOD AND THEIR RELATIVE SUITABILITY FOR JAPAN, by George Cawley, Esq. Read before the Asiatic Society of Japan, on 11th May, 1878.

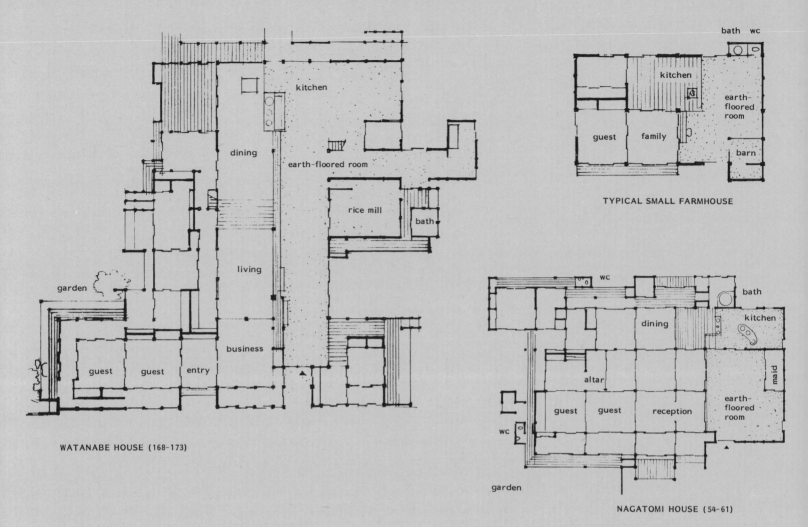

WATANABE HOUSE (168-173)

TYPICAL SMALL FARMHOUSE

NAGATOMI HOUSE (54-61)

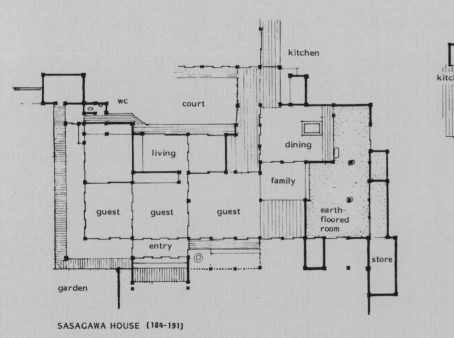

SASAGAWA HOUSE (184-191)

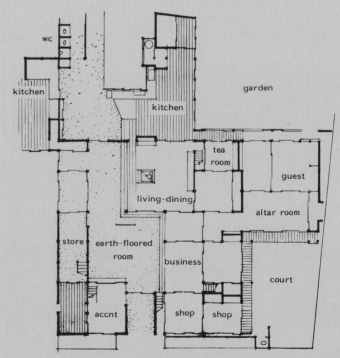

KUSAKABE HOUSE (145-151)

part of the house the little differentiation between the rooms, and the absence of fixed furniture allowed multiple use—entertaining, sitting, eating, and sleeping all took place on the tatami floor with cushions, low tables, and bedding stored in adjacent closets and brought out as needed.

These living rooms incorporated the interior posts into their dividing partitions, so a typical house was divided through the middle into four rooms. In more modest houses all rooms were used by both family and guests but in larger houses rooms overlooking the garden were kept strictly for guests.

Individual houses, even whole villages, whenever possible oriented south to capture the low winter sun (96). The important considerations of ventilation and view were easily accommodated by flexible window and door locations whatever the orientation. Most main entrances are on the side of the house parallel with the ridge. With few exceptions, the left portion of the house is the raised floor living space.

The master carpenter's basic design tool, made in consultation with the owner, was a plan drawn in ink on a board which outlined column positions and room relationships. From this simple diagram, making use of traditional methods and assumptions, he was able to project the final form of the house and all its structural complexities and details. The original of such a plan for the Kusakabe house in Takayama still survives and, for a house so marvelously complex, the stark simplicity of the plan is astounding (145-151).

Village and Urban Minka

In striking contrast to the densely built-up character of Mediterranean towns, is the low, dispersed character of most Japanese villages. Also, where the Italian, Iberian or Greek villages are usually perched high on hills beneath a castle or walled in for defense, in Japan with no history of foreign invasion, there is little of this defensive siting. I am not aware of any walled towns in Japan though some towns center on large castles. The siting of Japanese villages had much more to do with accessibility of agricultural land. The villages grew up near, but usually not on the best growing land, which in Japan's often rugged terrain, placed many villages at the base of hills on the edge of the lowland fields. Because of the limited space, the towns often became linear, stretching out along a road that followed the bottom of the hill (80-81).

Individual farmhouses standing among the fields are not a typical pattern, especially in the oldest parts of Japan around Kyoto and the west. Since few individual farmers owned their land until recently, the tenant farmers were gathered into villages and often traveled long distances to dispersed fields. Moreover, the cooperative nature of rice culture with complex irrigation systems of shared water and group planting and harvesting made the village essential.

Other than the common linear pattern, many villages developed randomly conforming to natural contours. Public spaces were few, consisting mainly of the grounds around the local shrine or temple. One elemental reason for the dispersed character of the Japanese village is that the typical minka form does not cluster easily and there is the real danger of spreading fire. Hence when compact villages

(continued on page 194)

TOHOKU
THE FAR COUNTRY

Tohoku, northeast Japan, was for much of its history the frontier—the far country well beyond the reach of established tradition. Remarkable local styles developed which reflected the region's rugged terrain, remoteness, severe climatic swings, and isolation due to poor communications and complex political control.

Although extensions of the basic folk style, houses in this region tend to be larger and more robust with huge extravagant roofs protecting against the heavy snows; they utilize enormous beams and columns reflecting the oversized roofs and abundance of timber; and they incorporate wide variations in room arrangements and construction details reflecting a freedom to innovate, unrestrained by the controls of the distant feudal government.

A town famous for the skill of its carpenters, Takayama contains several late 19th century folk-houses that are among the finest in Japan. Their style is reminiscent of Kyoto, as long ago outcasts from Kyoto settled here bringing with them the traditions of the capital.

Exposed structure was used with great effect in minka, but nowhere more dramatically than in the Kusakabe and Yoshijima houses by the famous carpenter-builder Jisuke Kawajiri of Takayama. His houses illustrate how the extraordinary skill and imagination of a local craftsman extends the limits of a folk tradition by exploiting its unexplored potentials. Kusakabe house (145-151).

As the home of a merchant, the interior of the Kusakabe house was divided and screened off for business as well as family use. The street elevation consists largely of grilled openings with inventive and expressive detailing. Originally of wood shingles, the roof was tiled in the 1920's to meet fire regulations (145-151).

Most minka interiors are rather dim since the only light must penetrate horizontally from beyond the wide overhangs. The light that floods these rooms from windows high in the gable walls comes as something of a shock. Enhanced by this unusual overhead light, the repetitive rhythm of the structure and the asymmetrical planes of the movable walls combine to create spaces with an astounding richness and sense of infinite extendibility—spaces that are among the most remarkable in Japanese architecture (148-153).

Kusakabe house (145-151)
The irregular size and placement of the beams required each member to be meticulously shaped and joined. Shrinkage cracks were reduced by adding deep grooves in the hidden face of beams and columns (151 top).
The unusual tree root suspends the kettle over the fire pit—an example of the Japanese delight in natural form (151 bottom).

This architecture developed within the structural cage as a series of patterns and planes of reference precisely defining certain portions of continuous space. (Yoshijima house 152, 153).

Not far from Takayama, in the mountain valleys of Hida, an unusual style of farmhouse evolved. It was large to accommodate the district's customary family size of 20 to 30 members under one roof and at the same time to provide attic space under the steep roof for raising silk worms. Now the silk business is gone and the children have escaped to the cities, so the huge houses are a burden for the few who remain.

The style is called Shirakawa or Gassho after villages in the Hida district. Shirakawa (154-155, 156).

Once rather numerous, many Shirakawa minka were destroyed or moved when a new dam flooded a major valley. Others have been abandoned or drastically remodeled.

The lower story or two were living quarters and the upper levels (157, 158) for trays of silk worms spread on latticed floors which allowed heat from below to rise. Mulberry leaves collected from the fields were fed to the worms (the sound of ten thousand munching silk worms is said to have made quite a din) producing cocoons from which the silk was extracted.

In much the same manner as the roofs of ancient houses, the huge Shirakawa roofs were constructed of poles lashed together with vines (157). The basic structure and roof were erected by groups of villagers, then carpenters from Takayama were called in to complete the interior details.

Shirakawa houses are superb examples of a form shaped by specialized needs, by available materials, and by local skills. Such specialized forms, however, become anachronisms when the conditions that produced them radically change.

Shirakawa (158), and Gassho-mura (159).

Always poor and at the bottom of a rigid medieval society, the peasants made do with materials they grew or could gather in the forests. Even when they could afford better they were forced to wear the simple, rough clothes of their station. Rain coats and hats were woven from straw and reeds. Occasionally still used in the 1960's, such things are now worn only during the special times of planting and harvesting, if then.

Tools, rain coats, baskets and carrying mats hanging under the eaves of a farm house near Shirakawa (160).

Farmhouses of Towa-cho near Tono display another of the varied roof styles of northern Honshu (161).

Two story houses between Takayama and Matsumoto (162 top, 163). One, a small shop, serves local farmers with a few canned goods, farm supplies, and some glass jars of candy for the children. It is fall and under the eaves hang persimmons, beets, and corn to dry (163).

In a scene from time immemorial, a woman sits near the fire sewing by light of the open door (162).

Barn near Matsumoto (164 top).
Tiny roadside shrines were once found throughout the countryside, especially at the site of some unusual natural form such as this tree trunk. Travelers and villagers made small offerings of food. Between Shirakawa and Matsumoto (164 bottom).
Around Matsumoto multi-storied houses sometimes presented their wide gable ends to the street permitting greater light and air for the upper floors. Matsumoto area (165).

People of the far country: a young girl, working at the rice harvest, wears a cloth mask to protect from the sharp stalks (166 top); rainy day wear (166 top right, bottom right); shrine festival, Shirakawa (166 bottom left); straw is bundled and then hung on racks (167 top left, bottom).

The bare racks erected in preparation for the harvest show the same primitive structural techniques of lashed poles and posts used in house construction since prehistoric times.

In the town of Sekigawa in Niigata are two large handsome minka of the landed gentry. The Watanabe house (168-173) built at the end of the 18th century fronts on the main village street. A facia board pegged to the rafters and shingles held in place by rows of stones are common details in the region.

As landowners and sake brewers, the owners required a large space for collecting, measuring and storing rice. This work room extends the full depth of the house. The kitchen and rice ovens are at the rear, well lighted by an expanse of shoji on the back wall. Behind the sliding wooden doors at the rear of the elevated floor are the family's private rooms and guest rooms. (170, 171).

Molded plaster and family crests decorate two of the several kura surrounding the Watanabe house (172 top).

The main entrance to the earth-floored room and a separate entrance to the business rooms just outside the heavy, decorated door on the right (172 bottom).

A surprising discovery in the Watanabe house is this elegant and spacious guest room overlooking a carefully manicured garden. Such a space makes clear why the Japanese endured the inherent drafts and chill of movable walls. For with all panels removed from beneath the great sweep of roof, garden and interior unite in a single, eloquent composition of geometry and nature (173).

The Sato house in Sekigawa, a product of the 19th century, has an unusual layered, thatched roof and a strong simple facade (174-176).

Sato house (176).
The pattern of shoji patched over the years with pages from an old ledger book (177).

Helmet-shaped roofs sheltering two and three story houses are common in the Gassan area, south of Tsuruoka (178 all).

Helmet-shaped roofs sheltering two and three story houses are common in the Gassan area, south of Tsuruoka (178 all).

The world famous potter, Shoji Hamada, was instrumental in preserving and popularizing many of Japan's folk traditions. As part of this interest, he acquired several old houses and other buildings from the district and reassembled them at his home in Mashiko. The Hamada gatehouse (179 bottom) and a workshop (179 top).

A large house at Hamada's, brought from near Nikko, clearly shows the increased scale and rugged proportions of northern houses in comparison with those of the Kyoto region. South facade and entrance (180 top, 181). Just inside the entrance (180 bottom).

Farmhouses near Tono. Some houses in this region are 'L' shaped in plan with the barn in one wing, the house in the other (182 bottom, 183).

At Mikata along the Sea of Japan the large Sasagawa house was formerly the official residence of a village mayor and land owner (whose descendants were still bitter over the distribution of their property after the war). Originally these large houses were in every sense working estates. The owners lived on the land, farming some land directly and supervising the peasants who lived on and worked the remainder. While life was certainly more secure for the landlords, they did not live in a very different style from their tenants, and their houses, though grander in scale, were built with the same materials and methods as all minka in the region.

Befitting its status, the Sasagawa house is surrounded by a wall entered through a gatehouse (184). There are the usual three entrances for different classes of visitors arranged in the conventional left to right order (185). One part of the middle entrance is screened off except for special festivities (186, 187). The patched shoji, while perhaps indicating the somewhat reduced circumstances of the present owners, adds a charming pattern to the facade. The wood shingle roof had just recently been replaced with a metal imitation.

The main pillars, daikoku-bashira, support large log beams fastened by wooden pegs. While apparently redundant and excessively complex, the many mortised joints compensated for the lack of diagonal bracing in the resistance of earthquakes—to say nothing of the impressive visual effect of which the owners are justifiably proud. The cutting of such large beams was strictly controlled in most of Japan and only those with status or connections were permitted to use them in their homes.

Since shoes were not worn on the wood or tatami floors, easily donned wooden geta were kept ready beside the steps (190).

The dining space at the rear of the earth-floored hall centers on a charcoal cooking pit over which hang iron kettles. Here, in strictly prescribed order, the family sat on the floor for meals or warmth while various foods were cooked before them and the hanging kettle heated water for tea. Except for the expansive tatami flooring, the scene was much the same in the more modest houses of the neighborhood (190, 191).

It is the great hall of this house, however, that is its most outstanding feature (188-191). Daylight from the oversized door and from the few windows on the side and back walls lightens the space even on this gray day.

EPITAPH

In the fall of 1986 I returned to Japan for the first time since the last of these photographs was made in 1964. I was well aware that Japanese society and economy had profoundly changed since those days, but I was unprepared for the visual devastation these changes had wrought.

The cities, of course, like most of the rest of the world, had long been beyond saving, but I had believed that the villages and the countryside, as in many parts of Europe and the Mediterranean, were timeless and that changes, if they occurred, would be minimal. I felt that, despite a few insidious intrusions (192), the Japanese tradition would remain eternally rooted in rural Japan.

Sadly, I found that such is not the case. Much of the Japan depicted in this book has either vanished or been irreparably altered over the last two decades.

Forced by vanishing skills, most of the few remaining farmhouses have thatch covered in metal (which at least preserves the form). Worse, countless ancient farmhouses have been ripped down and replaced by bland new houses that, except for size, would fit just as nicely in the suburban slums of Long Island or New Jersey. Shockingly, these modern houses offer no hint of the great Japanese building tradition—they are as ordinary and vapid as tract houses anywhere in the world.

The temples, gardens, palaces and other national treasures are still beautifully maintained, and there are occasional glimpses of unspoiled landscape.

Possibly naive, I had hoped that these great treasures and the apparent Japanese sensitivity to their environment combined with the strength, the modernity of Japanese architectural tradition would ease their transition into the 20th century and demonstrate how to make a world both contemporary and beautiful. Alas, it was not to be.

What happened to the famed Japanese sensibilities? Perhaps, as a friend who lives in Japan says, 'the Japanese still know what's beautiful—the problem is they don't know what's ugly.' Or, as I pointed out earlier, the devastation may stem from that remarkable Japanese ability to ignore the ugliness of the world just beyond their own tiny gardens.

Whatever the cause, the result is the wholesale destruction of a millennium of architectural and natural splendor in a mere twenty years—especially anguishing to one who spent so much time in its midst—as it must also be to many Japanese.

NFC, Kyoto 1986

A hint of the devastation soon to overtake the countryside is this addition in the Tamba countryside, circa 1964.

did expand, the houses were separated by small walled compounds not for defense, but for privacy, to discourage burglars, and for the prestige of an impressive gate (86). In larger towns such as Nara and Kyoto a special type of urban row house developed with dividing walls extending above the roof to minimize the ever-present threat of spreading fire (129).

Minka Today

In contrast with the folkhouses of the Mediterranean, where the undifferentiated spaces and forms seem infinitely adaptable to changing life styles, the specialized Japanese minka, as an integral part of a unique way of life, were doomed when that life changed. A few folkhouses will remain out of nostalgia—either preserved in museum settings or adopted by city families as weekend retreats where a few days of inconvenience are overcome by their charm.

Although their demise is inevitable, minka show the enormous power and beauty of building so thoroughly a part of a time and place—as well as the danger of overly specialized forms in changing times. In addition, as I have stressed again and again in this series, Japanese vernacular architecture should serve not as a source of particular forms, details, or techniques to be reproduced, but as a guide in understanding and reconnecting with the essential motivations that underlie architecture.

The village of Otsu on Lake Biwa along the pilgrimage route to Ise. The front sections of the houses, which open directly on the street, are shops and eating places for travellers. The artist of this 18th century travel guide has included many fascinating vignettes of daily life in the countryside.

BIBLIOGRAPHY AND NOTES

Akiyama, A. *Shinto and its Architecture*. Kyoto, 1936

6 Carver, Norman F., Jr. *Form and Space of Japanese Architecture*. Tokyo: Shokokusha, 1955

Cawley, George et al. *Transactions of the Asiatic Society of Japan*. Tokyo, 1878-1893

2 Dore, Ronald P. *Shinohata, Portrait of a Japanese Village*.
3 New York: Pantheon, 1978

Engel, Heinrich. *The Japanese House, A Tradition for Contemporary Architecture*. Tokyo: Tuttle 1964

Futagawa, Y., and Itoh, T. *The Essential Japanese House*. Tokyo: Weatherhill, 1967

Hokusai, *Manga (Sketch Book)*, vols 1-15, Tokyo: 1818-1878

Hall, John W. *Japan from Prehistory to Modern Times*. New York: Dell, 1970

4 Itoh, Teiji. *Traditional Domestic Architecture of Japan*. Tokyo: Weatherhill/Heibonsha, 1972

1 Morse, Edward S. *Japanese Homes and their Surroundings*. Salem, 1886. Reissued by Charles Tuttle Co., Tokyo 1972

Okakura-Kakuzo. *The Book of Tea*. Tokyo: Kenkyusha

Oto, Tokohiko. *Folklore in Japanese Life and Customs*. Tokyo: Kokusai Bunka Shinkokai, 1963

5 Paine, R.T., Soper, A. *The Art and Architecture of Japan*. Baltimore: Penguin Books, 1974

Tange, Kenzo, and Kawazoe, N. *Ise, Prototype of Japanese Architecture*. Cambridge: MIT Press, 1965

Watanabe, Y. *Shinto Art: Ise and Izumo Shrines*. Tokyo: Heibonsha, 1974

INDEX

Photographic pages indicated by ( ).

Ainu	9
Archetypes	10-14
Asia Society of Japan, quoted	141
Buddhist influence	7, 9
Chigi, roof projections	14
Chinese influences	9
Comfort, lack of	8
Daikoku-bashira, sacred column	140
Diagonal bracing, lack of	140
Earthquakes	8
Floors, earthen	138
Floors, tatami mats	138
Folkhouse and Japanese life	7
Folkhouse, as shelter	7
Folkhouse, as symbol	7, 46
Folkhouse, regulations	46
Folkhouses, misc.	42-48, 138-144, 194
Fusuma panels	138
Gassan area	(178)
Gassho-mura	(159)
Geku, Ise	(29)
Hamada House, Mashiko	(179-181)
Hida	(154-160)
Himeiji area	(49)
Hiroshima area	(90)
Hojo-ji, Tamba	(106)
Hokusai drawings	15, 44, 138
Horyu-ji (temple)	9
Horyu-ji village	(103)
Hozu-cho (Kyoto)	(82-83)
Ibaraki area	(86)
Ibogawa	(54-61)
Ichijima area	(89)
Ikoma, near Nara	(62, 80)
Ise Shrine, as archetype	11
Ise Shrines	11-14, (17-29, 40)
Ise, Geku Shrine	13, 16, (29)
Ise, Naiku Shrine	13, 16, (17-28, 40)
Ise, influence of	11-12
Ise, pilgrimage to	13, 194
Ise, rebuilding	13-14, 16, (24-25)
Ise, structural system	14
Isolation, regional	46-47
Itoh, Teiji, quoted	46
Izumo Shrine	(34-35)
Izumo Shrine, as archetype	11
Japan, climate	8
Japan, geography	8
Japan, history	9
Japan, isolation from world	7, 9
Japanese architecture, influence of	7
Japanese life, regimentation	44-45
Japanese people, origins	8, 9
Jimmu, emperor	9
Jisuke, Kawajiri, builder	47
Kameoka, near Kyoto	(53, 86)
Katsuogi, roof decorations	14
Katsura Palace	141
Keihoku, near Kyoto	(52, 88)
Kurashiki	(123-125)
Kusakabe House, Takayama	144, (145-151)
Kyoto	(78, 134)
Kyoto, Gion	(126-127, 131)
Kyoto, Nishijin	(128-129)
Kyoto, Ponte-cho	(130)
Kyoto, Shimabara	(132-133, 136)
Kyushu Island	(91)
Lake Biwa area	(75, 87)
Land ownership	45
Mashiko, near Nikko	(179-181)
Materials and construction	10, 47, 138
Mediterranean	45
Mikata	(184-191)
Minka	42-48, 138-144, 194
Minka, age of	43
Minka, defined	42
Minka, disappearance of	43
Minka, form and customs	45-46
Minka, general characteristics	42
Minka, history of	43
Minka, in village and town	143, 192
Minka, innovation	46-47
Minka, integration of	45-46
Miwa area	(89)
Miyama-cho	(93, 94-97, 101-104)
Morse, Edward S., quoted	7, 139
Nagatomi House, Ibogawa	(54-61)
Naiku (Ise)	(17-28, 40)
Nose	(50)
Omiya	(53)
Otsu	144
Pit houses	10
Pole houses	10
Privacy	8, 45
Raised houses, storehouses	10
Regional variations	46-47
Regional isolation	46-47
Rice culture	43-44 (72-78, 166-167)
Rice rack structures	10 (167)
Roof construction	15
Roof materials	139
Roof styles	138-139
Sadamitsu (Shikoku)	(98)
Saigoku, Southwestern Japan	48, (49-136)
Sasagawa House, Mikata	(184-191)
Sasayama	(79, 107)
Sato House, Sekigawa	(174-176)
Sekigawa	(168-176)
Semi-tropical style	8
Shijo, near Nara	(74)
Shimabara (Kyoto)	(132-133, 136)
Shinmei-gu, as archetype	11, (30-33)

Shinto influence	7
Shirakawa houses	47, (154-160)
Shitsukawa, Tamba	(109)
Shoin style	42
Shoji panels	47, 138
Shosoin, Nara, as archetype	11, (36-39)
Space and form	140-143
Space, flexible	140-143
Space, floor plans	141-143
Standardization	138
Stoicism	8
Structure and form	140
Structure, lack of diagonals	140
Sukiya style	42
Sumiya Inn, Kyoto	(132-133)
Takao, near Kyoto	(105)
Takayama houses	47, 144, (145-153)
Takehara	(122)
Tamba district	(50-109, 135, 192)
Tohara	(100)
Tohoku, Northeast Japan	144-192 (145-191)
Tojo village	(98)
Tombs, ancient mound	9
Tono area	(182-183)
Toro, ancient dwellings	10
Tsuruoka area	(178, 168-176)
Tsuruoka	42
Typhoons	8
Urban minka	194, (122-136, 145-153)
Variations, regional	46-47
Village life	44-45
Village patterns	143, 194
Village siting	143, 194
Wall construction	47, (103)
Watanabe House, Sekigawa	(168-173)
Wood construction	10
Yagi area, Tamba	(76-77)
Yoshijima House, Takayama	144, (152-153)
Yoshimura-tei House	(110-121)
Yoshitomi, Tamba	(53)

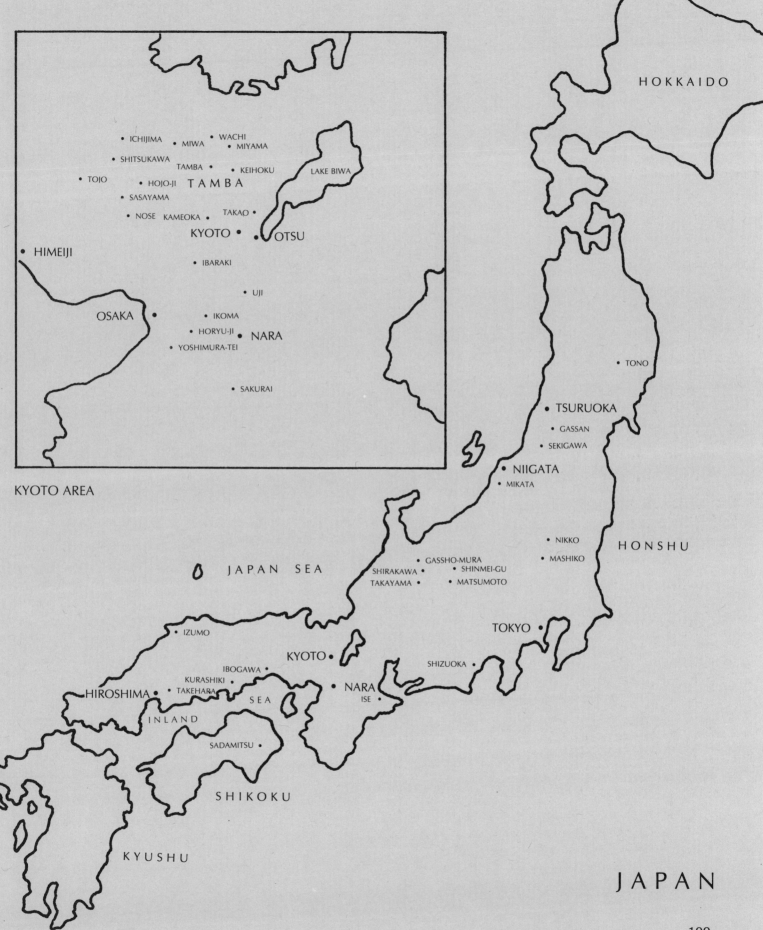

PHOTOGRAHIC NOTES

These photographs, taken during Fulbright grants to Japan in 1953-55 and 1964, are a distillation of thousands of images made when the villages and countryside were largely uncorrupted by the modern world. A great luxury in those more leisurely times was the opportunity to return again and again until the image on film was the same as the image in my head.

Rolleiflex and Hasselblad cameras were used, as well as several 35mm cameras. Films were developed and proofed in the kitchen of our Japanese house. Final prints were made recently on Ilford Multigrade and Kodak Polyfiber papers.

AVAILABILITY OF PRINTS:
Display prints of photographs in this or previous books are available, personally made by Norman F. Carver, Jr. Prices as of 1987 begin at $100 depending on size. Please write in care of the publisher, Documan Press Ltd., Box 387, Kalamazoo, MI 49005, USA.